YOUR KNOWLEDGE HAS VALUE

Bibliographic information published by the German National Library:

The German National Library lists this publication in the National Bibliography; detailed bibliographic data are available on the Internet at http://dnb.dnb.de .

Imprint:

Copyright © 2017 GRIN Verlag, Open Publishing GmbH
Print and binding: Books on Demand GmbH, Norderstedt Germany
ISBN: 9783668511682

This book at GRIN:

http://www.grin.com/en/e-book/369797/cholesterol-lowering-effect-of-tamarind-tamarindus-indica

Prem Jose Vazhacharickal, Jiby John Mathew, Sajeshkumar N.K., Berin Babu

Cholesterol lowering effect of Tamarind (Tamarindus indica)

GRIN Publishing

GRIN - Your knowledge has value

Since its foundation in 1998, GRIN has specialized in publishing academic texts by students, college teachers and other academics as e-book and printed book. The website www.grin.com is an ideal platform for presenting term papers, final papers, scientific essays, dissertations and specialist books.

Visit us on the internet:

http://www.grin.com/

http://www.facebook.com/grincom

http://www.twitter.com/grin_com

Cholesterol lowering effect of Tamarind (*Tamarindus indica*): a brief overview

Prem Jose Vazhacharickal, Jiby John Mathew, Sajeshkumar N.K and Berin Babu

ACKNOWLEDGEMENTS

Firstly we thank **God Almighty** whose blessing were always with us and helped us to complete this project work successfully.

We wish to thank our beloved Manager **Rev. Fr. Dr. George Njarakunnel,** Respected Principal **Dr. Joseph V.J,** Vice Principal **Fr. Joseph Allencheril,** Bursar **Shaji Augustine** and the Management for providing all the necessary facilities in carrying out the study. We express our sincere thanks to **Mr. Binoy A Mulanthra** (lab in charge, Department of Biotechnology) for the support. This research work will not be possible with the co-operation of many farmers.

Lastly, we extend our indebt thanks to patents, friends, and well wishers for their love and support.

Prem Jose Vazhacharickal*, Jiby John Mathew, Sajeshkumar N.K and Berin Babu

*Address for correspondence
Assistant Professor
Department of Biotechnology
Mar Augusthinose College
Ramapuram-686576
Kerala, India
premjosev@gmail.com

Table of contents

Table of contents...iii

Table of figures iv

Table of tables..v

List of abbreviations vi

Cholesterol lowering effect of Tamarind (*Tamarindus indica*): a brief overview 1

Abstract.. 1

1. Introduction 2

   1.1 Objectives 3
2. Review of literature.. 3

3. Hypothesis 7

4. Materials and Methods 7

   4.1 Study area... 7
   4.2 Collection of plant material... 7
   4.3 Preparation of *Tamarindus indica* fruit pulp extracts 7
   4.4 Phytochemical screening 7
   4.5 Preparation of cholesterol samples... 10
   4.7 Estimation of cholesterol 10
   4.8 Statistical analysis.. 10
5. Results and discussion... 17

6. Conclusions... 18

Acknowledgements 18

References.. 19

Table of figures

Figure 1. Map of Kerala showing the soil sample collection point. Authors own work
.... 11

Figure 2. Details of Tamarind (*Tamarindus indica*) a) fruits hanging on tree, b) harvested fruits, c) and e) fruits with outer skin removed and d) developing flower. Photo courtesy: Wikipedia. https://en.wikipedia.org/wiki/Tamarind. 12

Figure 3. Details of phytochemical analysis of Tamarind (*Tamarindus indicus*) fruit (water extract S1); A. terpenoids, B.leucoanthocyanins, C.flavonoids, D.proteins,E. alkaloids, F. anthraquinones, G. glycosides, H. coumarins, I. emodin J. carbohydrate, K. phenols, L. saponins, M. phlobatannins, N. steroids, O. anthocyanins. Authors own image. 13

Figure 4. Details of phytochemical analysis of Tamarind (*Tamarindus indicus*) fruit (water extract S2); A. terpenoids, B. leucoanthocyanins, C. flavonoids, D. proteins, E. alkaloids, F. anthraquinones, G. glycosides, H. coumarins, I. emodin J. carbohydrate, K. phenols, L. saponins, M. phlobatannins, N. steroids, O. anthocyanins. Authors own image. 14

Table of tables

Table 1. Chemical composition of *Tamarindus indica* fruit pulp. 6

Table 2. Preliminary phytochemical analysis of *Tamarindus indica* fruit extracts. 15

Table 3. Cholesterol estimation at different time intervals after treatment (n=3; values in mg/g sample)... 16

List of abbreviations

%	: Percentage
°C	: Degree celsius
μL	: Microlitre
CVD	: Cardio vascular disease
LDL	: Low density lipoprotein
TKP	: Tamarind kernel powder
VDL	: Very low density lipoprotein

Cholesterol lowering effect of Tamarind (*Tamarindus indica*): a brief overview

Prem Jose Vazhacharickal[1]*, Jiby John Mathew[1], Sajeshkumar N.K[1] and Berin Babu[1]

* premjosev@gmail.com

[1]Department of Biotechnology, Mar Augusthinose College, Ramapuram, Kerala, India-686576

Abstract

Aqueous extract of the fruit pulp of Tamarind (*Tamarindus indica*) were evaluated for cholesterol lowering effect, in vitro, against various fatty food materials. People consume food items made by chicken, beef, mutton, egg and fish and it contains large amount of fat. This study aims to analyze the effect of *Tamarindus indica* in reducing the cholesterol level in this fat compound using water extract of the pulp. For this fatty food samples like egg yolk, pork fat, chicken fat, ghee and cod liver oil were treated with the extract and cholesterol level was estimated by zak's method for a period of time. Phytochemical constituents present in water extract of *Tamarindus indica* pulp includes Alkaloids, saponins, steroids, phlobatannins, carbohydrate, terpenoids, phenols, coumarins and leucoanthocyanins. The in vitro cholesterol lowering effect of *Tamarindus indica* pulp extract shows a positive result on chicken fat, ghee and egg yolk. But in case of pork fat and cod liver oil no beneficial change was observed.

Keywords: Tamarind; Cholesterol; Zak's method.

1. Introduction

Tamarindus indica L. (Tamarind) belongs to the dicotyledonous family Fabaceae and subfamily caesalpiniaceae (Khanzada et al., 2008) known as Puli / Madhurappuli in Malayalam, has been used for centuries as a medicinal plant. Its fruits and leaves are the most valuable parts and have often used as a curative agent in several pharmacoeias. The leaves have an estimated protective activity associated with the presence of polyhydroxylated compounds with many of them flavonolic nature (Joyeux et al., 1995). Leaves also contain good levels of protein, fat, fiber and vitamins such as thiamine, riboflavin, niacin, ascorbic acid and β-carotene (El-Siddig et al., 2006).

A recent study demonstrated the hypolipidaemic effect of *Tamarindus indica* fruit pulp in hamsters. However, the biochemical and molecular mechanisms responsible for these effects have not been fully elucidated.

An increased level of lipids in the blood, including cholesterol and triglycerides, is known as Hyperlipidaemia. This increase is one of the significant risk factors involved in the development of cardiovascular disease (CVD) and diabetes (Pushparaj et al., 2007). CVD remains an important cause of mortality and morbidity worldwide (WHO, 2012). It is well recognized that increased levels of blood cholesterol particularly low density lipoprotein cholesterol (LDL-C) is an important risk factor for cardiovascular complications because it favours lipid deposition in blood vessels. Studies have clearly established that reduction of total cholesterol or LDLC is associated with decreased risk of atherosclerosis and coronary heart disease (Sovor et al., 2003).

The use of lipid-lowering drugs for the treatment of hyperlipidaemia lead to many side effect (Davidson et al., 2007). Therefore, there continues to be a high demand for new, more effective and less toxic oral hypolipidaemic drugs. Plant products are frequently considered to be less toxic and relatively free from side effects than synthetic drugs. So, plants play a major role in the introduction of new therapeutic agents and have received much attention as sources of biologically active substances including antioxidative, hypoglycaemic and hypolipidaemic agents.

Several studies have examined the pharmacological effect of *Tamarindus indica* L. The results showed that pectin of *Tamarindus indica* believed to have antioxidants that can reduce cholesterol and triglycerides and increase high density lipoprotein (HDL) (Chong et al., 2012). The pulp of Tamarind finds important place in chutneys, pickles, jams, curries, sauces, ice cream, sharbat and "tamarind fish". It is extensively used in

Kerala, Tamil Nadu, Karnataka and Andhra Pradesh cuisines, particularly in the preparation of Rassam and Sāmbhar. In these states of India many people were eating food product made up of meat and oil these leads to a high degree of CVD in these areas. They are consuming food items made by Chicken, Beef, Mutton, egg and fish and it contains large amount of fat. This study aims to analyse the effect of *Tamarindus indica* in reducing the cholesterol level in these fat compounds using aqueous extract of the pulp.

1.1 Objectives

The objectives of this study were to evaluate the phytochemical properties of water extract of *Tamarindus indica* L. fruit pulp and determine cholesterol lowering effect on various fatty food materials.

2. Review of literature

Plants have been used for medicinal purpose from prehistoric period. India has one of the richest plant medical traditions in the world. Herbal medicine, also called botanical medicine or phytomedicine, refers to using a plant's seeds, berries, roots, leaves, bark, or flowers for medicinal purposes. As herbs are natural products they are free from side effects, they are comparatively safe, eco-friendly and locally available. Medicinal plants are those plants which are rich in secondary metabolites and are potential source of drugs. These secondary metabolites include alkaloids, glycosides, coumarins, flavonoids, steroids etc. These chemicals work on the human body in exactly the same way as pharmaceutical drugs, so they can be beneficial and have harmful side effects just like conventional drugs. Scientific researchers in the recent past have come up to support the presence of medicinal activities in herbs recently, by carrying out research that can be found in the scientific literature (DeFeudis, 1991); these include herbs that produce an exceptional molecule to fight cancer and other diseases.

Cholesterol is a waxy substance in the blood plasma and in animal tissues. It is an organic compound ($C_{27}H_{460}$) in the steroid family and it is a white, crystalline substance that is odourless and tasteless in its pure state. Cholesterol is the most abundant sterol in animals; it is synthesized by the liver and consumed through diet. Cholesterol is essential for life serving as a building block for steroid hormones (testosterone and estrogen), and for cell walls. Generally in human body the liver

3

makes about cholesterol 80% and other 20% is comes from daily diet. Cholesterol is not required in the diet because the body can synthesize all it needs for essential functions (St. Jean, 2008). Foods containing cholesterol are foods from animals such as: egg yolks, meat, poultry, shellfish and whole milk and other dairy products and is poorly absorbed by the gut into the body.

Cholesterol cannot be dissolved in blood and are carried in the blood by particles called lipoproteins. And they are classified into Low-density Lipoprotein (LDL) and very low-density lipoprotein (VLDL) and High-density lipoproteins carrying cholesterol (HDL cholesterol) (Tabas, 2002). LDL cholesterol, also known as "bad cholesterol", is the major cholesterol carrier in the blood. HDL cholesterol (good cholesterol) on the other hand, carries about one forth blood cholesterol. A high level of cholesterol in the blood is a major risk for developing coronary heart disease which leads to heart attacks. Our total cholesterol should be under 200 mg/dL and anything over this number would be considered too high. Hypercholesterolemia is a condition when there is an extremely high level of cholesterol in the body. Cholesterol plays an important role in our nervous system i.e., in the formation of synapses and also in manufacture of steroids or cortisone like hormones, including vitamin D and sex hormones.

Tamarind is extensively used in the Indian system of medicine, Ayurveda. Scientifically known as *Tarmarindus indica* is a multipurpose tree of which almost every part is useful either nutritional or medicinal and a member of the family Fabaceae. Especially the fruit is beneficial; the sweetish acidic pulp of the fruit is a product of commercial importance (Kumar and Bhattacharya, 2008). It is extensively used in Tamil Nadu, Karnataka, Kerala and Andhra Pradesh cuisines, particularly in the preparation of Rassam and Sambhar.

The tree averages 20-25m in height and 1m in diameter, slow growing, but long lived, with an average life span of 80-200 years. Flowering generally occurs in synchrony with new leaf growth, which in most areas is during spring and summer (Orwa et al., 2009).The bark of Tamarind tree is brown gray coloured. It can tolerate diversity of soils like loam, sandy, clay soil, but well drained slightly acidic soil is best for its growth. The leaves are elliptical ovular, alternate, pinnate with reticulate venation and are a mass of bright green, dense foliage with feathery appearance. Normally, the leaves are evergreen but during hot season, they may be shed briefly in dry areas. Leaves

4

are 7.5-15 cm in length, each having 10 to 20 pairs of oblong leaflets (1.25-2.5 cm) and 5-6 mm wide. Leaves fold in cold damp weather and after sunset, due to the degradation of lupeol in dark, which is synthesized in light. Tamarind leaves are a fair source of vitamin C and ß-carotene and the mineral content is high, particularly potassium, phosphorous, calcium and magnesium (El-Siddig et al., 2006). Flowers are attractive pale yellow or pinkish, in small, lax spikes about 2.5cm in width. Flower buds completely enclosed by 2 bracteoles, which fall very early; sepals 4, petals 5, the upper 3 well developed, the lower 2 minute.

The seeds are 3-10, approximately 1.6cm long, irregularly shaped, and testa hard, shiny, and smooth. The major industrial product of tamarind seed is the tamarind kernel powder (TKP) which is an important sizing material used in the textile, paper, and jute industries (Kumar and Bhattacharya, 2008). Tamarind seed is also the raw material used in the manufacture of polysaccharide (jellose), adhesive and tannin. Tamarind seeds appear to be a good source of different mineral elements such as calcium, phosphorus, magnesium, sodium and potassium etc. Potassium is the element in highest concentration, with the values for the trace mineral copper also relatively high (Glew et al., 1997). The high concentration of potassium is nutritionally significant considering the fact that potassium plays a principal role in neuro-muscular function (Ajayi et al., 2006).

The fruit is a pod, indehiscent, sub cylindrical, 10-14 X 3 cm, straight or curved, velvety, rusty-brown; the shell of the pod is brittle. Tamarind is valued highly for its fruits, especially the pulp which is used for a wide variety of domestic and industrial purposes (Kulkarni et al. (1998) especially for food and beverages. Fruit pulp occurs as a reddish-brown, moist, sticky mass, in which yellowish-brown fibres and acidic with sweet taste and characteristic sour odour (Panara et al., 2014).

The most outstanding characteristic of tamarind is its sweet acidic taste, the acid due to mostly tartaric acid (El-Siddig et al., 1999). Tartaric is an unusual plant acid formed from the primary carbohydrate products of photosynthesis, and once formed, it is not metabolically used by the plant. The content of tartaric acid does not decrease during fruit ripening, as it is not utilized in fruit development. At the same time of fruit development; reducing sugars increase to 30-40% giving the sour fruit a sweeter taste. It is an excellent source of vitamin B and also contains minerals and exhibit high antioxidant capacity that appear to be associated with a high phenolic content, thus can be an important food source (El-Siddig et al., 2006).

Tamarind pulp typically contains:

Table 1. Chemical composition of *Tamarindus indica* fruit pulp.

Component	Concentration (%)
Water	20.6
Protein	3.1
Fat	0.4
Carbohydrate	70.8
Fibre	3.0
Ash	2.1

Thus, the pulp has low water content and a high level of protein, carbohydrates and minerals. It is a rich source of several macro and micro elements, including relatively high amounts of copper, manganese and zinc and also a good source of calcium and phosphorus (El-Siddig et al., 1999), but is unfortunately, extraordinarily low in iron (Glew et al., 2005). The consumption of 100 g tamarind fruit pulp by an adult will cover 10.69% of the recommended daily intake of calcium, 20.49% of magnesium, 14.21% of phosphorous, 12.07% of iron, 2.61% of manganese, 1.29% of zinc, 32.22% of copper and 9.21% of selenium, respectively (Almeida et al., 2009). Pulp was reported to have a high content of vitamin B (thiamine, riboflavin and niacin) as well as a small amount of carotene and vitamin C (El-Siddig et al., 1999; ICRAF, 2007).

Tamarindus indica leaves exhibited anti-emetic activity comparable to that of marketed medicine viz. chlorpromazine, Anti-histaminic activity. The pulp's extract exhibited remarkable antimicrobial activity against *Salmonella tymphimurium* and *Staphylococcus aureus* while possessed mild activity against *Aspergillus niger* (Doughari, 2006) and it also act as a Anti-measles and against mumps, Anti-diarrheal and Anti-dysentery agent. The fruits are used as preservatives. Seed extract reduces blood glucose levels and anti-venom the extract prolonged the clotting time moderately concomitantly. Tartaric acid present in tamarind act as powerful antioxidant that protects body from harmful free radicals, which can increase risk of cancer. Several other health benefits are wound healing activity, acaricidal activity, nerve health, treatment for malaria, fever, Helminthes infection, reduces acidity, nerve function, weight loss (Parle et al., 2012).

3. Hypothesis

The current research work is based on the following hypothesis

1) *Tamarindus indica* extracts are rich in various phytochemical components.

2) These extract could lower cholesterol levels.

4. Materials and Methods

4.1 Study area

Kerala state covers an area of 38,863 km^2 with a population density of 859 per km^2 and spread across 14 districts. The climate is characterized by tropical wet and dry with average annual rainfall amounts to 2,817 ± 406 mm and mean annual temperature is 26.8°C (averages from 1871-2005; Krishnakumar et al., 2009). Maximum rainfall occurs from June to September mainly due to South West Monsoon and temperatures are highest in May and November.

4.2 Collection of plant material

Tamarindus indica fruits were collected from Elanji village (survey no 529/1), Muvattupuzha Taluk, Ernakulum district, Kerala state and identified taxonomically and stored. These fruits were used for extract preparation. The extract was subjected to phytochemical analysis and its cholesterol lowering effect.

4.3 Preparation of *Tamarindus indica* fruit pulp extracts

Fruit pulp was separated from the seeds and air-dried. 2.5 g of fruits pulp were soaked in two separate conical flasks (S1 and S2) which contain water at room temperature for 24 h. The resulting extracts were then filtered and stored at 4°C.

4.4 Phytochemical screening

Chemical tests were carried out on the aqueous extract using standard procedures to identify the constituents as described by Sofowara (1993), Trease and Evans (1989) and Harborne (1973).

4.4.1 Test for alkaloids

7

Two ml of plant extract was taken in a test tube and few drops of Hager's reagent were added. Yellow precipitate shows positive result for alkaloids.

4.4.2 Test for anthraquinones

Three ml of plant extract was taken in a test tube and three ml of benzene and five ml of ten percentage NH_3 were added. Formation of pink, violet or red coloration in ammonical layer detect the presence of anthraquinones.

4.4.3 Test for anthocyanins

Two ml of plant extract was taken in a test tube and two ml of 2N HCl and NH_3 were added. Formation of pinkish red to bluish violet coloration indicates the presence of anthocyanins.

4.4.4 Test for carbohydrate

Two ml of plant extract was taken in a test tube and ten ml of water, two drops of twenty percentage ethanolic α naphthol and two ml of conc.H_2SO_4 were added. Formation of reddish violet ring at the junction shows the presence of carbohydrates.

4.4.5 Test for coumarins

Two ml of extract was taken in a test tube and three ml of ten percentage NaOH was added. Formation of yellow colour gives positive result to coumarins.

4.4.6 Test for emodins

Two ml of plant extract was taken in a test tube and two ml of NH_4OH and three ml of benzene were added. Formation of red colour indicates the presence of emodins.

4.4.7 Test for flavonoids

Five ml of dilute ammonia solution were added to a portion of the plant extract followed by addition of concentrated H_2SO_4. A yellow colouration observed in each extract indicated the presence of flavonoids. The yellow colouration disappeared on standing.

4.4.8. Test for glycosides

Two ml of plant extract was taken in a test tube and two ml of chloroform and two ml of acetic acid were added. Formation of violet to blue to green coloration shows the presence of glycosides.

4.4.9 Test for leucoanthocyanins

Five ml of isoamyl alcohol taken in a test tube and five ml of plant extract was added. Turn organic layer into red detects the presence of leucoanthocyanins.

4.4.10 Test for phlobatannins

Deposition of a red precipitate when an extract of each plant sample was boiled with one percentage aqueous hydrochloric acid was taken as evidence for the presence of phlobatannins.

4.4.11 Test for proteins

One ml of plant extract was mixed with one ml of conc.H_2SO_4 in a test tube. Formation of white precipitate indicate the presence of proteins

4.4.12 Test for phenols

Few ml of the plant extract was taken in attest tube and few ml of lead acetate was added to it. Formation of white precipitate detects the presence of phenols.

4.4.13 Test for saponins

Ten ml of the extract was mixed with five ml of distilled water and shaken vigorously for a stable persistent froth. The frothing was mixed with three drops of olive oil and shaken vigorously, then observed for the formation of emulsion.

4.4.14 Test for steroids

Two ml of extract was taken in a test tube and two ml chloroform and two ml of conc.H_2SO_4 was added. Formation of reddish brown ring at the junction shows the presence of steroids.

4.4.15Test for terpenoids

Five ml of each extract was mixed in two ml of chloroform, and concentrated H_2SO_4 (three ml) was carefully added to form a layer. A reddish brown colouration of the inter face was formed to show positive results for the presence of terpenoids.

4.5 Preparation of cholesterol samples

One gram of the sample was dissolved in one ml of chloroform and stored in brown bottle for further use (Varley, 2004).

4.6 Treatment

200 µl of extract was added to each of the sample prepared and mixed well. These are used for the periodic (24 hour interval) determination of cholesterol.

4.7 Estimation of cholesterol

The amount of cholesterol in each sample was estimated by Zak's method before and after treatment (Varley, 2004).

4.8 Statistical analysis

The survey results were analyzed and descriptive statistics were done using SPSS 12.0 (SPSS Inc., an IBM Company, Chicago, USA) and graphs were generated using Sigma Plot 7 (Systat Software Inc., Chicago, USA).

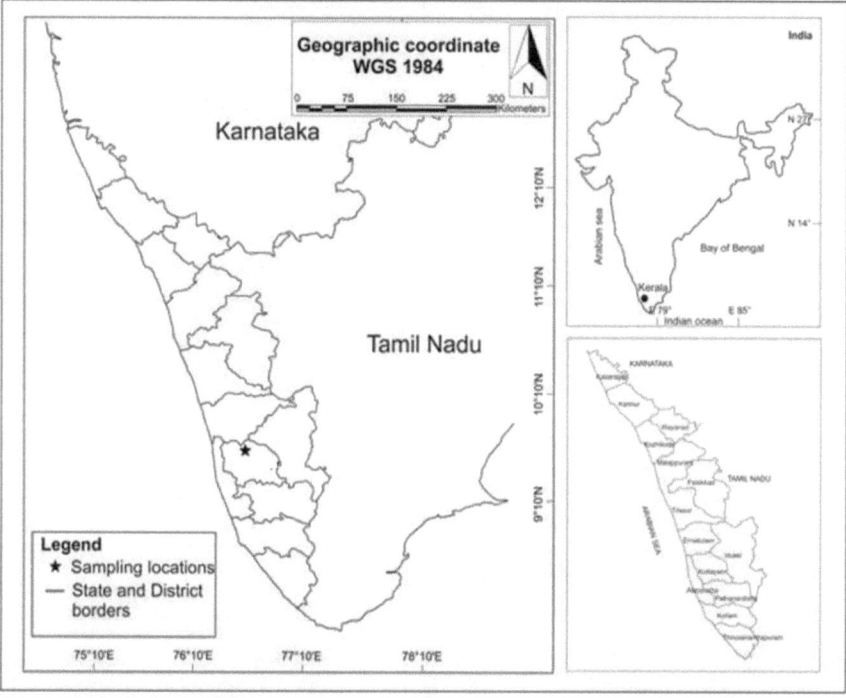

Figure 1. Map of Kerala showing the soil sample collection point. Authors own work

Figure 2. Details of Tamarind (*Tamarindus indica*) a) fruits hanging on tree, b) harvested fruits, c) and e) fruits with outer skin removed and d) developing flower. Photo courtesy: Wikipedia. https://en.wikipedia.org/wiki/Tamarind.

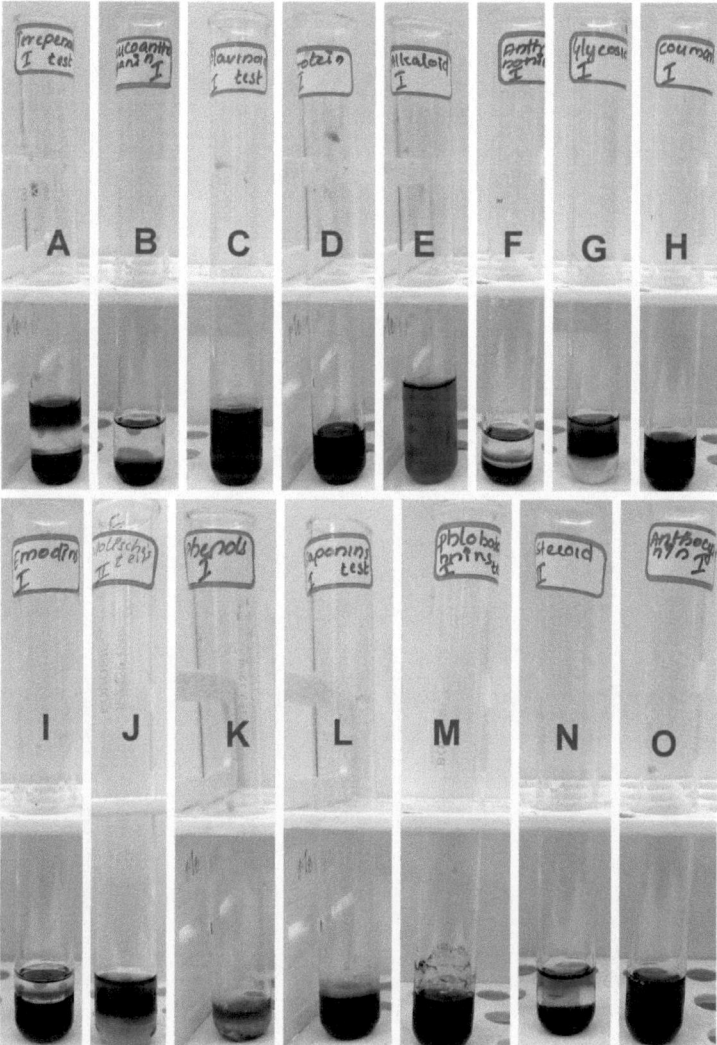

Figure 3. Details of phytochemical analysis of Tamarind (*Tamarindus indicus*) fruit (water extract S1); A. terpenoids, B.leucoanthocyanins, C.flavonoids, D.proteins,E. alkaloids, F. anthraquinones, G. glycosides, H. coumarins, I. emodin J. carbohydrate, K. phenols, L. saponins, M. phlobatannins, N. steroids,   O. anthocyanins. Authors own image.

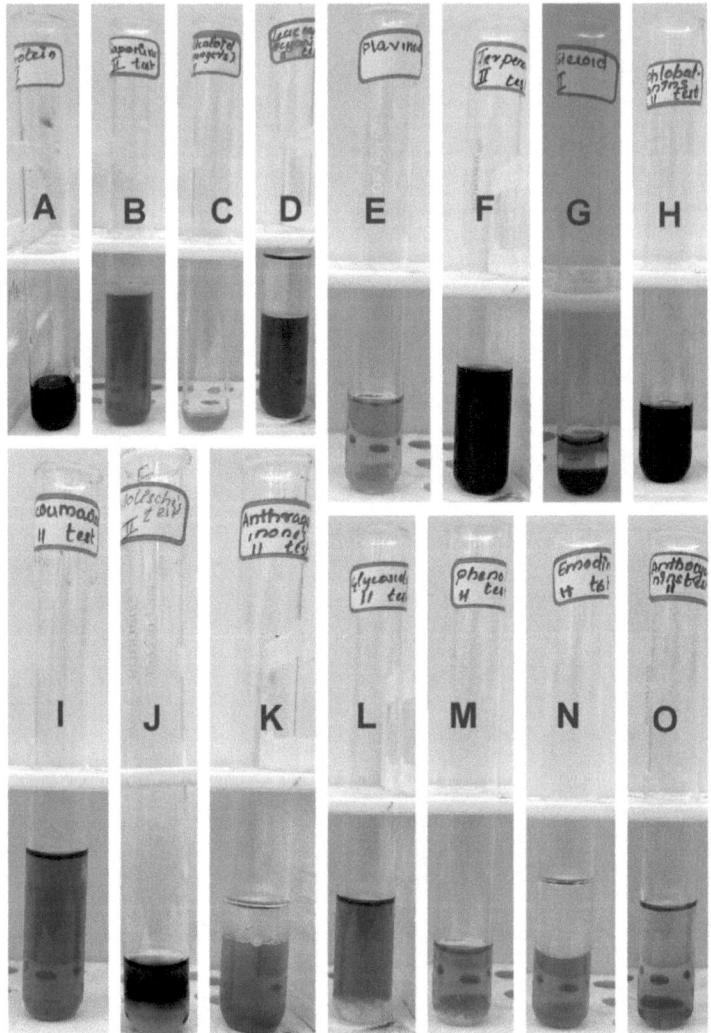

Figure 4. Details of phytochemical analysis of Tamarind (*Tamarindus indicus*) fruit (water extract S2); A. terpenoids, B. leucoanthocyanins, C. flavonoids, D. proteins, E. alkaloids, F. anthraquinones, G. glycosides, H. coumarins, I. emodin J. carbohydrate, K. phenols, L. saponins, M. phlobatannins, N. steroids, O. anthocyanins. Authors own image.

Table 2. Preliminary phytochemical analysis of *Tamarindus indica* fruit extracts.

Sl.No	Phytoconstituents	Aqueous Extract(S1)	Aqueous Extract(S2)
1.	Alkaloid	+	+
2.	Anthocyanins	-	-
3.	Anthraquionones	-	-
4.	Carbohydrates	+	+
5.	Coumarins	+	+
6.	Emodins	-	-
7.	Flavonoids	-	-
8.	Glycoside	-	-
9.	Leucoanthocyanins	+	+
10.	Phenols	+	+
11.	Phlobatannins	+	+
12.	Protein	+	+
13.	Saponins	+	+
14.	Steroids	+	+
15.	Terpenoids	+	+

+ indicates presence of phytochemicals
- indicates absence of phytochemicals

Table 3. Cholesterol estimation at different time intervals after treatment (n=3; values in mg/g sample).

Day	Substrate					
	Egg yolk	Pork Fat	Chicken Fat	Ghee	Cod liver oil
1	0.027 ± 0.01	0.013 ± 0.01	0.027 ± 0.03	0.024 ± 0.02	0.013 ± 0.03
2	0.027 ± 0.01	0.019 ± 0.01	0.021 ± 0.02	0.021 ± 0.01	0.013 ± 0.01
3	0.024 ± 0.01	0.019 ± 0.02	0.021 ± 0.01	0.019 ± 0.01	0.013 ± 0.01
4	0.019 ± 0.01	0.018 ± 0.01	0.021 ± 0.01	0.016 ± 0.01	0.013 ± 0.01
5	0.019 ± 0.01	0.011 ± 0.00	0.019 ± 0.01	0.016 ± 0.01	0.012 ± 0.00
6	0.016 ± 0.02	0.011 ± 0.01	0.019 ± 0.01	0.016 ± 0.01	0.012 ± 0.00
7	0.016 ± 0.01	0.011 ± 0.00	0.019 ± 0.01	0.016 ± 0.01	0.012 ± 0.00
8	0.016 ± 0.02	0.010 ± 0.01	0.019 ± 0.01	0.013 ± 0.01	0.012 ± 0.01
9	0.016 ± 0.00	0.010 ± 0.01	0.013 ± 0.00	0.013 ± 0.01	0.012 ± 0.01
10	0.016 ± 0.01	0.010 ± 0.01	0.013 ± 0.01	0.013 ± 0.01	0.012 ± 0.01
11	0.013 ± 0.01	0.010 ± 0.00	0.013 ± 0.00	0.011 ± 0.01	0.012 ± 0.01
12	0.013 ± 0.01	0.009 ± 0.00	0.011 ± 0.00	0.011 ± 0.00	0.012 ± 0.00
13	0.013 ± 0.01	0.009 ± 0.00	0.006 ± 0.00	0.008 ± 0.00	0.012 ± 0.00

Numbers represent means ± one standard error (SE) of mean.

5. Results and discussion

Phytochemical constituents present in water extract of *Tamarindus indica* L. pulp includes Alkaloids, saponins, steroids, phlobatannins, carbohydrate, terpenoids, phenols, coumarins and leucoanthocyanins. The presence of some secondary metabolites like alkaloids, reducing sugars, steroids and saponins was in agreement with the works of Doughari, (2006). However, the finding of terpenoids, phenols, coumarins and leucoanthocyanins were not in line with that of Ahmad and Abdul (2014). The chemical composition of amino acids, fatty acids, and minerals of tamarind plant parts have been reported. Differences in values found in the literature are likely to be due to differences in genetic strains, stages of maturity at which the plant parts were harvested, growing conditions (Glew et al., 2005), harvesting and handling techniques as well as to differences in analytical methodologies.

The general population was advised to limit cholesterol intake to less than 300 mg/day (Adult Treatment Panel III, 2002; Krauss et al., 2001). A single egg yolk contains approximately 215 mg to 275 mg of cholesterol. A single egg yolk thus exceeds the recommended daily intake of cholesterol.

The in vitro cholesterol lowering effect of *Tamarindus indica* L. pulp extract shows a positive result on chicken fat, ghee and egg yolk. But in case of pork fat and cod liver oil beneficial change was not observed.

In plants, these compounds function to attract beneficial and repel harmful organisms, serve as photoprotectants and respond to environmental changes. In humans, they can have complementary and overlapping actions, including antioxidant effects, modulation of detoxification enzymes, stimulation of the immune system, reduction of inflammation, modulation of steroid metabolism and antibacterial and antiviral effects. The presence of coumarins lowered cholesterol and increased high-density lipoprotein level.

Martinello et al. (2006) found that the pulp of fruit of *Tamarindus indica* L. is able to reduce the levels of total cholesterol (50%), non-HDL cholesterol (73%), and triglycerides (60%). Research by Jindal et al. (2011) stated the extract of tamarind pulp can decrease total cholesterol and triglycerides significantly.

The water extract of *Tamarindus indica* L. fruit flesh exert effect for lowering total cholesterol and triglyceride on different fatty food materials under study. Further research need to be carried out to determine the toxicity/safety level of the plant extracts, administration as well as to isolate and identify the active compound responsible for the activity.

6. Conclusions

The study was conducted to analyse the phytochemical characteristics of *Tamarindus indica* and evaluates its cholesterol lowering activity. The plant parts have benefited mankind in various ways, including treatment of many diseases. Most of the parts like leaves, bark, flowers, fruits, seeds, roots or the whole plant are used as alternative medicine to treat a variety of diseases. The fruits and leaves are the most valuable parts and have often used as a curative agent in several pharmacopeias.

Phytochemicals are biologically active, naturally occurring chemical compounds found in plants, which provide health benefits. Phytochemical tests were carried out on the aqueous extract and standard procedures used to identify the constituents. Various phytochemical compounds like alkaloids, flavonoids, saponins, glycosides, emodins, proteins, carbohydrate, terpenoids, tannins, coumarins and phenols were found in the fruit extracts of the plant.

Five common fatty food materials like egg yolk, pork fat, chicken fat, ghee and cod liver oil was mixed with the fruit extract of *Tamarindus indica* and the cholesterol level in each was evaluated in vitro, after and before treatment. The in vitro analysis shows that cholesterol level of fatty food materials treated with the extract show reducing day by day except pork fat and cod liver oil. The change in cholesterol level was not evident in the early days but significant reduction was observed after a week.

Acknowledgements
The authors are grateful for the cooperation of the management of Mar Augusthinose college for necessary support. Technical assistance from Binoy A Mulanthra is also acknowledged.

References

Adult Treatment Panel III. Third Report of the National Cholesterol Education Program (NCEP) Expert Panel on Detection, Evaluation, and Treatment of High Blood Cholesterol in Adults (Adult Treatment Panel III) final report. Circulation. 2002;106:3143–421.

Ajayi, I. A., Oderinde, R. A., Kajogbola, D. O., & Uponi, J. I. (2006). Oil content and fatty acid composition of some underutilized legumes from Nigeria. *Food Chemistry*, *99*(1), 115-120.

Almeida, M. M. B., Sousa, P. H. M. D., Fonseca, M. L., Magalhães, C. E. C., Lopes, M. D. F. G., & Lemos, T. L. G. D. (2009). Evaluation of macro and micro-mineral content in tropical fruits cultivated in the northeast of Brazil. *Food Science and Technology (Campinas)*, *29*(3), 581-586.

Choi, W. S., Chang, S. H., Kim, J. E., & Lee, S. E. (2013). Hypolipidemic effects of scoparone and its coumarin analogues in hyperlipidemia rats induced by high fat diet. *Journal of the Korean Society for Applied Biological Chemistry*, *56*(6), 647-653.

Chong, U. R. W., Abdul-Rahman, P. S., Abdul-Aziz, A., Hashim, O. H., & Junit, S. M. (2012). Tamarindus indica extract alters release of alpha enolase, apolipoprotein AI, transthyretin and Rab GDP dissociation inhibitor beta from HepG2 cells. *PloS one*, *7*(6), e39476.

Davidson, M. H., & Robinson, J. G. (2007). Safety of aggressive lipid management. *Journal of the American College of Cardiology*, *49*(17), 1753-1762.

DeFeudis, F. V. (1991). *Ginkgo biloba extract (EGb 761): Pharmacological activities and clinical applications*. Elsevier.

Doughari, J. H. (2006). Antimicrobial activity of Tamarindus indica Linn. *Tropical Journal of Pharmaceutical Research*, *5*(2), 597-603.

El-Siddig, K., Gunasena, H. P. M., Prasad, B. A., Pushpakumara, D. K. N. G., Ramana, K. V. R., Vijayanand, P., & Williams, J. T. (2006). Tamarind. *Tamarindus indica L. International Centre for Underutilised Crops, University of Southampton, SO17 1BJ, UK, Southampton*.

El-Siddig, K., Williams, J. T., Gunasena, H. P. M., Prasad, B. A., Pushpakumara, D. K. N. G., Ramana, K. V. R., & Vijayanand, P. (2006). Tamarind fruits for the future. *Tamarind: Tamarindus indica*.

Glew, R. H., Vanderjagt, D. J., Lockett, C., Grivetti, L. E., Smith, G. C., Pastuszyn, A., & Millson, M. (1997). Amino acid, fatty acid, and mineral composition of 24 indigenous plants of Burkina Faso. *Journal of food composition and analysis*, *10*(3), 205-217.

Glew, R. S., Vanderjagt, D. J., Chuang, L. T., Huang, Y. S., Millson, M., & Glew, R. H. (2005). Nutrient content of four edible wild plants from West Africa. *Plant Foods for Human Nutrition*, *60*(4), 187-193.

Harborne, J. B. (1973). Phenolic compounds. In *Phytochemical methods* (pp. 33-88). Springer Netherlands.

Hu, J., Nakatani, M., Lalusin, A. M., Kuranouchi, T. & Fujimura, T. (2003). Genetic anlysis of sweet potaoto using RAPD markers in Kerala. *Breeding Science*, 53(*1*), 297-304.

ICRAF (2007). Agroforestry Tree Database: *Tamarindus indica* L., World Agroforestry Centre. ICRAF. URL http://www.worldagroforestrycentre.org (visited on 30.01.2017).

Jindal, V., Dhingra, D., Sharma, S., Parle, M., & Harna, R. K. (2011). Hypolipidemic and weight reducing activity of the ethanolic extract of Tamarindus indica fruit pulp in cafeteria diet-and sulpiride-induced obese rats. *Journal of Pharmacology and Pharmacotherapeutics*, *2*(2), 80-84.

Joyeux, M., Lobstein, A., Anton, R., & Mortier, F. (1995). Comparative antilipoperoxidant, antinecrotic and scavanging properties of terpenes and biflavones from Ginkgo and some flavonoids. *Planta Medica*, *61*(02), 126-129.

Joyeux, M., Mortier, F., & Fleurentin, J. (1995). Screening of antiradical, antilipoperoxidant and hepatoprotective effects of nine plant extracts used in Caribbean folk medicine. *Phytotherapy Research*, *9*(3), 228-230.

Kapur, M. A., & John, S. A. (2014). Antimicrobial Activity of Ethanolic Bark Extract of *Tamarindus indica* against some Pathogenic Microorganisms. *International Journal of Current Microbiology and Applied sciences*, *3*(3), 589-593.

Khanzada, S. K., Shaikh, W., Sofia, S., Kazi, T. G., Usmanghani, K., Kabir, A., & Sheerazi, T. H. (2008). Chemical constituents of Tamarindus indica L. medicinal plant in Sindh. *Pak. J. Bot*, *40*(6), 2553-2559.

Khanzada, S. K., Shaikh, W., Sofia, S., Kazi, T. G., Usmanghani, K., Kabir, A., & Sheerazi, T. H. (2008). Chemical constituents of Tamarindus indica L. medicinal plant in Sindh. *Pak. J. Bot, 40*(6), 2553-2559.

Krauss, R. M., Eckel, R. H., Howard, B., Appel, L. J., Daniels, S. R., Deckelbaum, R. J., ... & Lichtenstein, A. H. (2001). Revision 2000: a statement for healthcare professionals from the Nutrition Committee of the American Heart Association. *The Journal of nutrition, 131*(1), 132-146.

Krishnakumar, K. N., Rao, G. P., & Gopakumar, C. S. (2009). Rainfall trends in twentieth century over Kerala, India. *Atmospheric environment, 43*(11), 1940-1944.

Kulkarni, D., Dwivedi, A. K., & Singh, S. (1998). Performance evaluation of tamarind seed polyose as a binder and in sustained release formulations of low drug loading. *Indian Journal of Pharmaceutical Sciences, 60*(1), 50.

Kulkarni, R. S., Gangaprasad, S., & Swamy, G. S. K. (1993). Tamarind: economically an important minor forest produce. *MFP News, 3*(3).

Kumar, C. S., & Bhattacharya, S. (2008). Tamarind seed: properties, processing and utilization. *Critical Reviews in Food Science and Nutrition, 48*(1), 1-20.

Martinello, F., Soares, S. M., Franco, J. J., Santos, A. C., Sugohara, A., Garcia, S. B., & Uyemura, S. A. (2006). Hypolipemic and antioxidant activities from *Tamarindus indica* L. pulp fruit extract in hypercholesterolemic hamsters. *Food and Chemical Toxicology, 44*(6), 810-818.

Orwa, C., Mutua, A., Kindt, R., Jamnadass, R. & Simons, A. 2009. Agroforestry Database: a tree reference and selection guide, version 4.0. World Agroforestry Centre, Kenya.

Panara, K., Harisha, C. R., & Shukla, V. J. (2014). Pharmacognostic and Phytochemical evaluation of fruit pulp of Tamarindus Indica linn. *International Journal of Ayurvedic Medicine, 5*(1). 37-42.

Parle, M., & Dhamija, I. (2012). Anxiolytic potential of *Tamarindus indica*. *Annals of Pharmacy and Pharmaceutical Sciences, 3*(2), 67-71.

Pushparaj, P. N., Low, H. K., Manikandan, J., Tan, B. K. H., & Tan, C. H. (2007). Anti-diabetic effects of Cichorium intybus in streptozotocin-induced diabetic rats. *Journal of Ethnopharmacology, 111*(2), 430-434.

Sever, P. S., Dahlöf, B., Poulter, N. R., Wedel, H., Beevers, G., Caulfield, M., & Mehlsen, J. (2003). Prevention of coronary and stroke events with atorvastatin in hypertensive patients who have average or lower-than-average cholesterol concentrations, in the Anglo-Scandinavian Cardiac Outcomes Trial-Lipid Lowering Arm (ASCOT-LLA): a multicentre randomised controlled trial. *The Lancet, 361*(9364), 1149-1158.

Shankaracharya, N. B. (1998). Tamarind-Chemistry, technology and uses: A critical appraisal. *Journal of Food Science and Technology, 35*(3), 193-208.

Sofowora, A. (1993). Recent trends in research into African medicinal plants. *Journal of Ethnopharmacology, 38*(2-3), 197-208.

Souza-Fagundes, E. M., Queiroz, A. B., Martins Filho, O. A., Gazzinelli, G., Corrêa-Oliveira, R., Alves, T., & Zani, C. L. (2002). Screening and fractionation of plant extracts with antiproliferative activity on human peripheral blood mononuclear cells. *Memorias do Instituto Oswaldo Cruz,97*(8), 1207-1212.

St Jean, N. (2008). Lowering Cholesterol: through the use of plant sterols and stanols. Senor Honors Project, Paper 82, University of Rhode Island.

St Jean, N. (2008). Lowering Cholesterol: through the use of plant sterols and stanols.

Tabas, I. (2002). Cholesterol in health and disease. The Journal of clinical investigation, 110(5), 583-590.

Trease, G. E., & Evans, W. C. Pharmacognosy. 1989. *Bailliere Tindall, London*, 45-50.

Varley, H. (2004). Practical clinical Biochemistry, 4th edition, Heinemann Medical, UK.

World Health Organization. (2012). Alcohol in the European Union: consumption, harm and policy approaches: Final report, Copenhagen.